June 2013

DEFENSE FORENSICS

Additional Planning and Oversight Needed to Establish an Enduring Expeditionary Forensic Capability

Highlights of GAO-13-447, a report to congressional requesters

Why GAO Did This Study

DOD used expeditionary forensics for collecting fingerprints and deoxyribonucleic acid (DNA) to identify, target, and disrupt terrorists and enemy combatants in Iraq and Afghanistan. The increased incidence of improvised explosive devices and other asymmetric threats has increased demand for expeditionary forensic capabilities. Many of DOD's expeditionary forensic activities are resourced through DOD's Overseas Contingency Operations funds. DOD estimates that it cost between $800 million and $1 billion of these funds from 2005 through 2012 to support expeditionary forensics activities in Iraq and Afghanistan. However, as military operations are projected to draw down in Afghanistan, this funding is expected to substantially decline by the end of 2014. Consequently, DOD is taking steps to establish expeditionary forensics as an enduring capability in DOD's base budget. GAO was asked to examine DOD's expeditionary forensic capability. This report assessed the extent to which DOD has taken steps to establish an enduring expeditionary forensic capability. To address this objective, GAO reviewed relevant policy, plans, and budget estimates, and interviewed cognizant DOD officials.

What GAO Recommends

GAO is making four recommendations to DOD, including incorporating key elements in its strategic plan, periodically reviewing and evaluating DOD components' proposed forensic budget estimates—including expeditionary forensics, and issuing guidance on collecting and reporting forensic budget data. DOD concurred with all four recommendations.

View GAO-13-447. For more information, contact Brian Lepore at (202) 512-4523 or leporeb@gao.gov

What GAO Found

The Department of Defense (DOD) has taken some important steps to establish an enduring expeditionary forensic capability by issuing a concept of operations in 2008, followed by a directive in 2011 to establish policy and assign responsibilities. As required by the directive, DOD has drafted a strategic plan to guide the activities of the Defense Forensic Enterprise, including expeditionary forensics. Although the plan includes a mission statement, and goals and objectives—two of the five key elements identified by GAO as integral to a well-developed strategic plan—it does not identify approaches for how goals and objectives will be achieved, milestones and metrics to gauge progress, and resources needed to achieve goals and objectives. GAO's prior work has shown that organizations need a well-developed strategic plan to identify and achieve their goals and objectives effectively and efficiently. Officials in the Office of the Under Secretary of Defense for Acquisition, Technology, and Logistics (OUSD(AT&L)) said that they decided to create a concise, high-level strategic plan and that they plan to issue guidance tasking the DOD components to develop individual implementation plans that include milestones. However, approaches, metrics, and resources needed to accomplish its goals and objectives were absent from the draft guidance. GAO discussed this omission with OUSD(AT&L), and in response, this office plans to revise its draft guidance. Also, the forensic strategic plan has been in draft for 2 years having undergone multiple revisions, and is still undergoing DOD internal review with no publication date set, and by extension, a publication date has not been set for the proposed DOD component implementation plans. The lack of an approved strategic plan and associated implementation plans limits DOD's ability to prioritize its efforts to develop an enduring expeditionary forensic capability by the end of 2014.

Moreover, OUSD(AT&L) has not reviewed and evaluated the adequacy of DOD components' expeditionary forensic budget estimates for fiscal years 2013 through 2018, as required by DOD's directive. OUSD(AT&L) officials said that they were waiting for the DOD components to finalize their budget estimates for fiscal years 2013 through 2018, and waiting for the Joint Capabilities Integration Development System to validate their forensic requirements. Regardless, reviewing and evaluating the DOD components' proposed budget estimates allows OUSD(AT&L) to advise the DOD components on their resource allocation decisions with respect to expeditionary forensic capabilities. OUSD(AT&L) officials cited several factors that also affected their ability to review and evaluate the DOD components' forensic budget data, such as aggregation of components' forensic budget estimates with other costs. Moreover, these officials said the directive does not provide guidance to DOD components on how to collect and report forensic budget data. GAO's Standards for Internal Control in the Federal Government notes that agencies should provide policy and guidance to determine the effectiveness and efficiency of operations. Until OUSD(AT&L) reviews and evaluates the adequacy of DOD components' forensic budget estimates, and guidance is in place to inform forensic budget collection and reporting, OUSD(AT&L) will continue to experience challenges with identifying the costs associated with DOD's expeditionary forensic capabilities.

United States Government Accountability Office

Contents

Letter		1
	Background	4
	DOD Has Taken Steps to Establish an Enduring Expeditionary Forensic Capability, but Additional Actions Are Needed	9
	Conclusions	15
	Recommendations for Executive Action	16
	Agency Comments and Our Evaluation	16
Appendix I	Scope and Methodology	20
Appendix II	Comments from the Department of Defense	23
Appendix III	GAO Contact and Staff Acknowledgments	26
Related GAO Products		27

Tables

	Table 1: GAO's Assessment of DOD's Draft Forensic Strategic Plan	10
	Table 2: DOD Organizations Contacted	20

Figure

	Figure 1: Examples of Expeditionary Forensic Laboratory Modules	5

Abbreviations

DNA	deoxyribonucleic acid
DOD	Department of Defense
OUSD(AT&L)	Office of the Under Secretary of Defense for Acquisition, Technology, and Logistics
USD(AT&L)	Under Secretary of Defense for Acquisition, Technology, and Logistics

This is a work of the U.S. government and is not subject to copyright protection in the United States. The published product may be reproduced and distributed in its entirety without further permission from GAO. However, because this work may contain copyrighted images or other material, permission from the copyright holder may be necessary if you wish to reproduce this material separately.

441 G St. N.W.
Washington, DC 20548

June 27, 2013

Congressional Requesters

U.S. military forces have used expeditionary forensics to identify, target, disrupt, and detain terrorists and enemy combatants in recent and ongoing conflicts in Iraq and Afghanistan, and to support host nation rule of law and capacity building in those areas and others such as the Horn of Africa. DOD defines forensics as the application of multi-disciplinary scientific processes to establish facts that may be used for the collection, identification, and comparison of materials such as deoxyribonucleic acid (DNA) and latent fingerprints.[1] For the purposes of this report, expeditionary forensics refers to the employment of forensic applications by an armed force to accomplish a specific objective in a foreign country. For example, in 2012 the Department of Defense (DOD) provided forensic support to about 120 Afghan court cases linking latent fingerprints and DNA evidence to enemy combatants, resulting in a 97 percent conviction rate. DOD has traditionally used forensics for law enforcement and medical purposes, such as identifying and prosecuting criminals and determining the identification of human remains. However, DOD's concept of operations notes that the increased incidence of improvised explosive devices and other asymmetric threats that U.S. military forces have encountered has created an increased demand for expeditionary forensic capabilities across the full range of military operations.[2] According to a 2011 Army Audit Agency report, the conflicts in Iraq and Afghanistan have helped revolutionize the department's use of expeditionary forensics in general, and latent fingerprints and DNA in particular,[3] and a 2012 DOD Office of the Inspector General report noted that U.S. forces in Afghanistan have used latent fingerprints and DNA to

---

[1] Latent fingerprints are images left on a surface touched by a person.

[2] DOD, *Capstone Concept of Operations for Department of Defense Forensics* (Washington, D.C.: July 18, 2008).

[3] U.S. Army Audit Agency, *Workforce Requirements for Expeditionary Forensics*, Audit Report No. A-2012-0031-FFD (Alexandria, VA: December 27, 2011) (For Official Use Only).

link known enemy combatants to captured enemy material such as improvised explosive devices.[4]

Many of DOD's expeditionary forensic activities are currently resourced through DOD's Overseas Contingency Operations funds—appropriations provided by Congress outside of the department's base appropriations process. The Office of the Under Secretary of Defense for Acquisition, Technology, and Logistics (OUSD(AT&L)), said that from 2005 through 2012, DOD estimated that it cost between $800 million and $1 billion in Overseas Contingency Operations appropriations to support expeditionary forensic activities in Iraq and Afghanistan. These funds covered such capabilities as the Army's Expeditionary Forensic Laboratories, the Navy's Combined Explosives Exploitation Cells, and Special Operations Command's expeditionary forensic activities. However, as U.S. military operations are projected to draw down in Afghanistan by the end of 2014,[5] Overseas Contingency Operations funding for expeditionary forensics is expected to decline substantially. While DOD maintains base funding for traditional forensic applications and Special Operations Command's sensitive site exploitation program,[6] DOD officials said that base funding levels are not adequate to cover the additional costs of current and emerging expeditionary forensic activities. Moreover, Army officials noted that it would cost considerably more in time and resources to recreate the Army's expeditionary forensic capability to support any future mission if current capabilities are allowed to expire. Consequently, DOD is taking steps to establish expeditionary forensics as an enduring capability prior to the projected drawdown of operations in Afghanistan by the end of 2014.

To date, we have issued four reports related to DOD biometrics (e.g., the measurement and analysis of an individual's unique physical or behavioral characteristics that can be used to verify personal identity) and forensics. (These reports are listed in the Related GAO Products section

---

[4]Department of Defense Inspector General, *Semiannual Report to the Congress* (Washington, DC.: April 1, 2012 – September 30, 2012).

[5]GAO, *Afghanistan: Key Oversight Issues*, GAO-13-218SP (Washington, D.C.: February 11, 2013).

[6]The Special Operations Command established a base funding account to cover a majority of its sensitive site exploitation program, which includes support for expeditionary forensic activities.

at the end of this report.) Building on our body of work examining DOD biometrics and forensics, as requested, we examined DOD's expeditionary forensic capability. Specifically, this report addresses the extent to which DOD has taken steps to establish an enduring expeditionary forensic capability.

We narrowed the scope of our review to laboratory applications of expeditionary forensics, and focused on latent fingerprints and DNA —the two most prevalent forensic disciplines U.S. forces have relied on in Iraq and Afghanistan. Concurrently, OUSD(AT&L) contracted with CNA's Center for Naval Analyses to assess the department's overall forensic activities, to include expeditionary forensics, and issue a report for the Secretary of Defense by the summer of 2013. We met with officials from OUSD(AT&L) and CNA's Center for Naval Analyses to ensure our work did not overlap with this assessment.

To assess the steps DOD has taken to establish an enduring expeditionary forensic capability, we analyzed relevant policy and guidance that describe DOD's plans to organize, train, and equip forces to sustain expeditionary forensic activities. We also interviewed planning, operations, and resource management officials from the Office of the Secretary of Defense, each of the four military services, all six geographic combatant commands, and Special Operations Command to discuss the status of planned and ongoing initiatives related to expeditionary forensics. We reviewed and analyzed DOD's forensic strategic plan[7] to determine if it included key strategic planning elements that are consistent with our previous work on developing and implementing strategic plans.

We also reviewed and analyzed the military services' and Special Operations Command's current and projected forensic budget data to determine if the budget data were sufficiently reliable and met the department's requirements in the DOD forensic directive.[8] We assessed the reliability of the budget data by interviewing knowledgeable officials and reviewing related documentation and written responses to our

---

[7]Office of the Under Secretary of Defense for Acquisition, Technology, and Logistics, *DOD Forensic Enterprise Draft Strategic Plan* (as of February 2013).

[8]Department of Defense, *DOD Forensic Enterprise,* DOD Directive 5205.15E (April 26, 2011).

questions on data reliability. We identified several issues concerning the reliability of the budget data obtained from OUSD(AT&L), the military services, and Special Operations Command, including the sources from which the budget data were derived, the consistency in how the budget data were compiled, and the manner in which the budget data were verified. As a result, we determined that the budget data were not sufficiently reliable. Therefore, we are making a recommendation that addresses OUSD(AT&L)'s ability to review and evaluate the DOD components' forensic budget data by calling for the development of budget collecting and reporting guidance.

More detailed information on our scope and methodology can be found in appendix I of this report. We conducted this performance audit from May 2012 through June 2013 in accordance with generally accepted government auditing standards. Those standards require that we plan and perform the audit to obtain sufficient, appropriate evidence to provide a reasonable basis for our findings and conclusions based on our audit objectives. We believe that the evidence obtained provides a reasonable basis for our findings and conclusions based on our audit objectives.

Background

DOD's Expeditionary Forensic Capabilities

Improvised explosive devices have caused numerous injuries and fatalities among U.S. and coalition forces while carrying out operations in Iraq and Afghanistan. To mitigate this threat, DOD has taken a number of actions, such as deploying its expeditionary forensic capabilities. In 2003, DOD established the Combined Explosives Exploitation Cells to provide technical intelligence on improvised explosive devices and render safe these devices and other combustible materials so that they can be forensically analyzed to obtain, among other things, latent fingerprints of the individuals responsible for manufacturing and placing the devices. In 2006, DOD further expanded its use of expeditionary forensics by establishing the joint expeditionary forensic facilities to analyze materials, such as ammunition and clothing items collected on the battlefield, to help identify enemy combatants through latent prints and DNA analysis, among other things. In 2011, DOD consolidated the Combined Explosives Exploitation Cells and the joint expeditionary forensic facilities to form Expeditionary Forensic Laboratories under the purview of the Army in order to realize efficiencies and minimize redundancies. These combined laboratories, like their predecessors, are modular, deployable, containerized units that can operate 24 hours a day, 7 days a week, and provide the capability to forensically analyze material such as latent fingerprints, DNA, explosives, drugs, and firearms and tool marks in response to warfighter needs. Figure 1 shows three of the Expeditionary Forensic Laboratory modules operating in Afghanistan.

Figure 1: Examples of Expeditionary Forensic Laboratory Modules

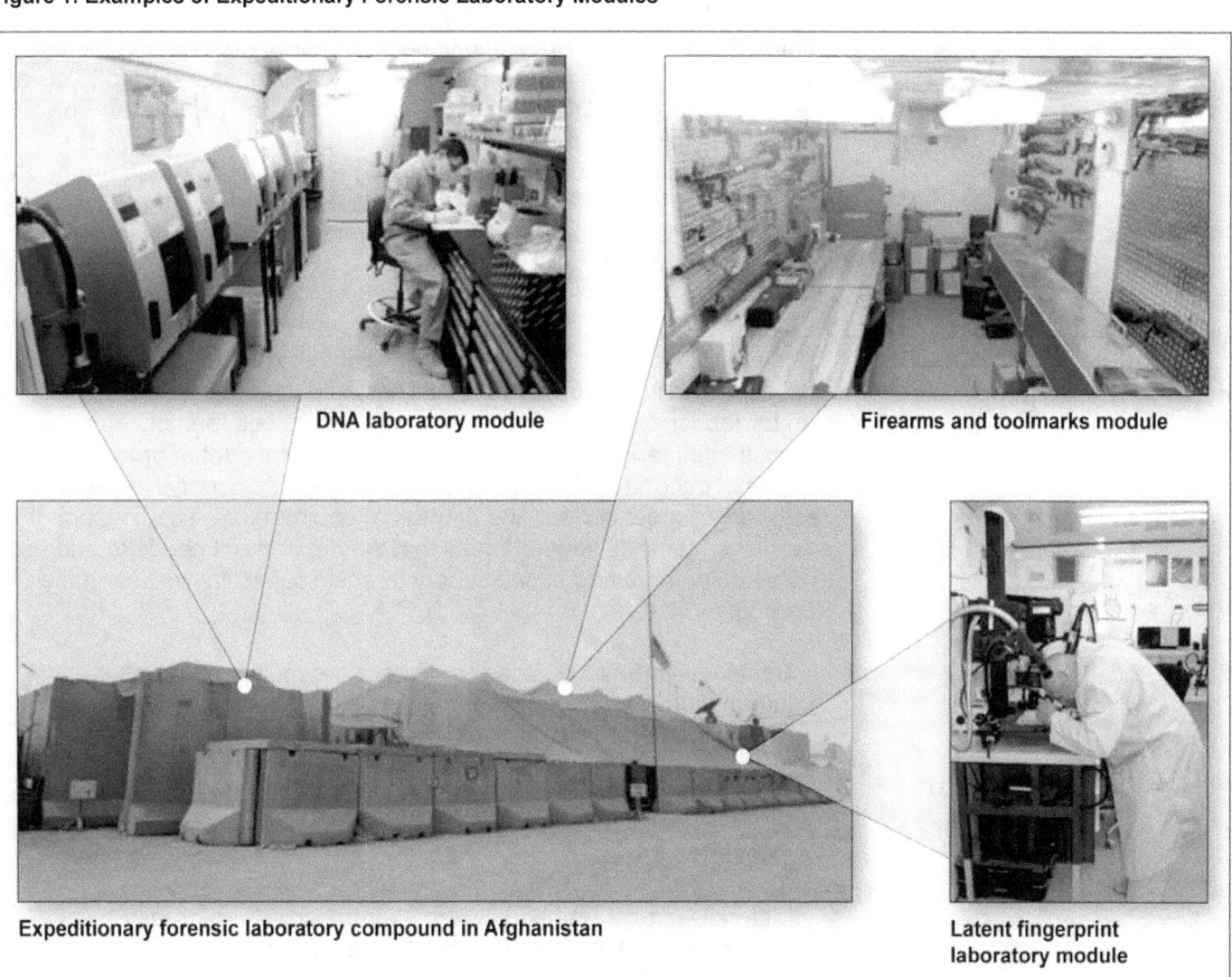

Source: U.S. Army Criminal Investigation Laboratory.

The Army also established a reachback operations center to oversee the deployment of Expeditionary Forensic Laboratories and to provide forensic expertise and analytical capabilities to process any overflow of forensic cases from Iraq and Afghanistan. In addition to the Army's expeditionary forensic capabilities, the Navy has provided staff support to the Combined Explosives Exploitation Cells through its Explosive Ordnance Disposal Technical Division. The Marine Corps has relied on the Army's Expeditionary Forensic Laboratories to forensically analyze material that the Marine Corps has collected on the battlefield for subsequent targeting and prosecutions. The Special Operations Command has developed expeditionary forensic toolkits and exploitation analysis centers as part of its sensitive site exploitation program to collect latent fingerprint and DNA samples, among other things. The Air Force focuses on digital and multimedia forensic applications to support operations such as counterintelligence to process, analyze, and translate data collected from electronic devices. We did not include the Air Force in our review because we focused on the forensic disciplines of latent fingerprints and DNA.

DOD Policy Governing Its Forensic Enterprise

In 2008, DOD issued a concept of operations that identifies existing, emerging, and future forensic capabilities, and calls for the department to plan for robust, fully-coordinated, and well-resourced forensic applications across the full range of military operations. The concept of operations states that integrating forensics—particularly expeditionary forensics—is necessary to meet current and emerging requirements. The concept of operations identifies several areas that the department needs to address to develop an enduring expeditionary forensic capability, including the following:

- Doctrine—forensic doctrine to address the multiple uses of information, varying timelines, and scientific challenges across the strategic, operational, and tactical levels;

- Training, Leadership and Education—forensic training and education for all levels of operations, and to ensure leadership understands the value of forensics;

- Materiel—equipment and systems for the collection, transfer, exploitation, dissemination, and storage of forensic material and information; and

- Facilities—deployed and institutional forensic laboratories to meet mission requirements.

In an effort to address the areas identified in the concept of operations, in 2009, DOD initiated a capabilities based assessment, which includes validating expeditionary forensic requirements as an enduring capability within the department, through the Joint Capabilities Integration and Development System. This system guides the development of capabilities from a joint perspective, to help identify capability gaps and validate the requirements of proposed capability solutions to mitigate those gaps. As part of this process, from 2011 to 2013, DOD developed and validated an initial capabilities document that contained information on those capabilities needed to support current and future forensic activities across the department—including expeditionary forensics. Furthermore, in early 2013, DOD initiated a change recommendation[9] that will lay out specified required forensic capabilities and projected costs. The process to review and approve the change recommendation is scheduled to begin in January 2014.

In April 2011, DOD issued a directive that, among other things, established policy and assigned responsibilities within the department to develop and maintain an enduring, holistic, forensic capability to support the full range of military operations—including law enforcement, medical, intelligence, and expeditionary forensics.[10] DOD refers to this holistic effort as the Defense Forensic Enterprise. The directive assigns roles and responsibilities for the following key DOD entities:

- USD(AT&L) is the principal staff assistant for the Defense Forensic Enterprise and Chair of the Forensic Executive Committee. The USD(AT&L) is responsible for, among other things, coordinating and publishing a Defense Forensic Enterprise strategic plan, and reviewing the adequacy of forensic-related acquisition programs and associated budget submissions to ensure they meet the Enterprise's program requirements and objectives.[11]

- The Secretary of the Army is designated as the DOD Executive Agent for Forensics for disciplines relating to, among other things, DNA and

---

[9]Also referred to as a Doctrine, Organization, Training, Materiel, Leadership and Education, Personnel, Facilities and Policy Change Recommendation.

[10]Department of Defense, DOD Directive 5205.15E. The directive does not cover unique forensic applications that support technical nuclear or technical chemical and biological disciplines, and specialized intelligence collected through reconnaissance programs.

[11]In October, 2011, USD(AT&L) appointed a Director, Defense Biometrics and Forensics, to oversee his responsibilities for the Defense Forensics Enterprise.

latent fingerprints.[12] The Secretary of the Army also is responsible for coordinating with the DOD components[13] to program for and budget sufficient resources to support common forensic requirements, and designating a DOD Center of Excellence to promote collaboration of best practices for forensic capabilities.

- The Secretary of the Air Force is designated as the DOD Executive Agent for Digital and Multimedia Forensics relating to computer and electronic devices, audio analysis, image analysis, and video analysis. The Secretary of the Air Force is responsible for coordinating with the DOD components to program and budget sufficient resources for digital and multimedia forensics, and designating a Center of Excellence for these forensic disciplines.[14]

- DOD components are required to support various programs and policies related to the Defense Forensic Enterprise, such as consulting and coordinating with USD(AT&L) on the establishment of forensic programs and policies; coordinating and integrating strategies, concepts, capabilities, and requirements to prevent unnecessary duplication of forensic activities; and formulating and executing budgets for forensic activities.

- The Chairman of the Joint Chiefs of Staff is required to coordinate combatant commanders' forensic requirements with the DOD Executive Agent for Forensics across the full range of military operations. The Chairman of the Joint Chiefs of Staff is also required to develop operational joint doctrine related to forensic capabilities,

---

[12]The Secretary of the Army also is responsible for forensic disciplines relating to serology, questioned documents, drugs, trace materials, firearms and tool marks, as well as forensic medicine disciplines such as forensic pathology, forensic anthropology, and forensic toxicology.

[13]DOD's forensic directive defines DOD components as the Office of the Secretary of Defense, Office of the Chairman of the Joint Chiefs of Staff and the Joint Staff, the Office of the Inspector General of the Department of Defense, military departments, combatant commands, defense agencies, field activities, and all other organizational entities within the department. Based on the scope of our review, for the purposes of this report, the phrase "DOD components" refers only to the Army, Navy, Marine Corps, and Special Operations Command.

[14]We did not examine digital and multimedia forensics because it is outside the scope of our review.

validate joint requirements for forensic capabilities for the joint force, and coordinate theater-specific requirements for forensic capabilities.

DOD Has Taken Steps to Establish an Enduring Expeditionary Forensic Capability, but Additional Actions Are Needed

DOD has taken some important steps to establish an enduring expeditionary forensic capability by issuing a concept of operations, and a directive that calls for a strategic plan addressing DOD's enterprise-wide forensics, including expeditionary forensics, but has not completed its strategic plan or reviewed and evaluated the adequacy of DOD components' budget estimates. DOD has drafted a forensic strategic plan;[15] however, it does not include three of the five key elements identified by GAO as integral to a well-developed strategic plan. Specifically, the plan does not include approaches for how the goals and objectives will be achieved, milestones and metrics to gauge progress, and resources needed to achieve the goals and objectives. Also, the forensic strategic plan has been in draft for 2 years without a scheduled completion date. OUSD(AT&L) officials stated that the draft is still undergoing internal review within the department. DOD plans to capture some of the elements of a strategic plan in implementation plans to be developed by DOD components; however, the timeframe for issuance of these implementation plans is unknown since they will follow publication of the strategic plan. In addition, according to DOD's forensic directive, USD(AT&L) is to review the adequacy of DOD components' forensic-related acquisition programs and associated budget submissions to ensure they meet DOD's overarching forensic requirements and objectives. However, at the time of our review, OUSD(AT&L) had not reviewed and evaluated forensic budget estimates for fiscal years 2013 through 2018 to ensure they meet the Defense Forensic Enterprise requirements and objectives. OUSD(AT&L) officials said that they had difficulty identifying forensic activities that are not specifically cited within the DOD components' forensic budget estimates. Further, OUSD(AT&L) officials said that while the DOD directive calls for OUSD(AT&L) to conduct a review of forensic-related programs and budget submissions, it does not provide guidance on how forensic budget data should be collected and reported by the DOD components.

---

[15]Office of the Under Secretary of Defense for Acquisition, Technology, and Logistics, *Forensic Enterprise Draft Strategic Plan* (as of February 2013).

DOD Has Issued Guidance but Its Draft Strategic Plan Does Not Include Some Key Elements

DOD has taken some important steps to establish an expeditionary forensic capability by issuing a concept of operations in 2008 and a directive in 2011 to establish policy and assign responsibilities. Consistent with the directive, DOD has drafted a strategic plan to guide the activities of the Defense Forensic Enterprise, including expeditionary forensics. However, DOD's strategic plan does not include three of the five key elements identified by GAO as integral to a well-developed strategic plan. The plan includes a mission statement and goals and objectives, but does not include approaches for how these goals and objectives will be achieved, milestones and metrics to gauge progress, and resources (e.g., funding and personnel) needed to achieve these goals and objectives. GAO's prior work on strategic planning has shown that organizations need a well-developed strategic plan to identify and achieve their goals and objectives effectively and efficiently.[16] Table 1, below, lists the key elements of a strategic plan and indicates those that are and are not included in DOD's draft strategic plan for the Defense Forensic Enterprise.

Table 1: GAO's Assessment of DOD's Draft Forensic Strategic Plan

Key elements of a strategic plan	Key elements included
Mission statement	Yes
Goals and objectives	Yes
Approaches for accomplishing goals and objectives	No
Milestones and metrics to gauge progress	No
Resources needed to meet goals and objectives	No

Source: GAO analysis of DOD data.

Note: Data are from DOD's draft strategic plan for the Defense Forensic Enterprise.

Consistent with these criteria, the draft strategic plan includes a mission statement and four broad goals that outline DOD's intent to meet the department's overarching needs for the Defense Forensic Enterprise. These goals are to (1) provide forensic information that is accurate and

---

[16]GAO, *Defense Logistics: A Completed Comprehensive Strategy Is Needed to Guide DOD's In-Transit Visibility Efforts*, GAO-13-201 (Washington, D.C.: February 28, 2013); GAO, *Depot Maintenance: Improved Strategic Planning Needed to Ensure That Army and Marine Corps Depots Can Meet Future Maintenance Requirements*, GAO-09-865 (Washington, D.C.: September 17, 2009); and GAO, *Results-Oriented Cultures: Implementation Steps to Assist Mergers and Organizational Transformations*, GAO-03-669 (Washington, D.C.: July 2, 2003).

timely, (2) develop cost-effective methods for providing forensic capabilities, (3) maximize the availability and accessibility of forensic-related information, and (4) invest in forensic research and technology. The strategic plan also includes a number of subordinate objectives that are linked to the four goals. However, the strategic plan does not include some key elements, such as approaches for how the objectives will be achieved, milestones and metrics to gauge DOD's progress, and the resources needed to meet its goals and objectives. For example, the two objectives under the plan's third goal—to maximize the availability and accessibility of forensic-related information—call for creating and promoting a forensic information-sharing culture that supports multiple users within DOD as well as with interagency and international partners. However, these objectives neither describe an approach for accomplishing them, nor include milestones, metrics, and resources. Without these key elements, DOD will be unable to measure its progress and adjust its approach when warranted, and identify the resources necessary to achieve the goals and objectives outlined in the strategic plan. Consequently, DOD may not have the information it needs to make well-informed decisions about forensics, including setting priorities for expeditionary forensic capabilities in an increasingly constrained fiscal environment.

The April 2011 DOD directive assigned OUSD(AT&L) responsibility for coordinating and publishing the DOD enterprisewide forensics strategic plan. An OUSD(AT&L) official stated that his office decided to create a concise, high-level strategic plan that included broad goals and subordinate objectives, but not milestones. According to this official, this decision was consistent with several other DOD strategic documents including the National Defense Strategy. OUSD(AT&L) officials said that after the strategic plan is issued, their office plans to issue guidance that will task the DOD components to develop individual implementation plans that include milestones. However, neither the draft strategic plan nor the proposed implementation plans would include approaches for how the goals and objectives will be achieved, metrics to gauge DOD's progress, or the resources needed to accomplish its goals and objectives. Based on our observations of the draft guidance, OUSD(AT&L) is revising its guidance to direct the DOD components to include approaches and metrics, in addition to milestones, in their proposed implementation plans. OUSD(AT&L) officials explained that resource information will continue to be omitted because OUSD(AT&L) does not have the authority to direct the DOD components on how to allocate their resources. However, OUSD(AT&L) can advise DOD components' resourcing decisions in a manner consistent with the goals and objectives articulated in the Defense Forensic Enterprise strategic plan.

The concept of operations, forensic directive, and strategic plan are important actions DOD has taken to establish expeditionary forensics as an enduring capability since 2008. However, the forensic strategic plan has undergone multiple revisions and has been in draft for 2 years. An OUSD(AT&L) official said that a publication date has not been set for the strategic plan, and by extension, a publication date has not been set for the proposed DOD component implementation plans. OUSD(AT&L) officials stated that the draft is still undergoing internal review within the department. GAO's *Standards for Internal Control in the Federal Government* notes that organizations must have relevant, reliable, and timely information to achieve their goals and objectives.[17] Otherwise, their ability to achieve their goals and objectives can be adversely affected.

In the absence of an approved forensic strategic plan, and in accordance with DOD's forensic directive, several of the military services and combatant commands have been independently developing guidance and plans to address their specific expeditionary forensic needs as shown in the following examples.

- The Army, as the DOD Executive Agent for Forensics, is developing guidance that includes a process for identifying and prioritizing DOD's expeditionary forensic requirements and capabilities that are common to all of the DOD components.

- The Marine Corps issued an identity operations strategy that includes a discussion of expeditionary forensic force structure and equipment considerations.[18]

- The Special Operations Command established a program with funding and issued guidance on sensitive site exploitation that includes training and education, and equipment such as forensic

---

[17]GAO, *Standards for Internal Control in the Federal Government,* GAO/AIMD-00-21.3.1 (Washington, D.C.: November 1999).

[18]U.S. Marine Corps, *U.S. Marine Corps Identity Operations Strategy 2020 Implementation Plan* (August 14, 2012). The Navy has not developed guidance on expeditionary forensics, but Navy officials have stated that they plan to develop a strategy that mirrors the Marine Corps identity operations strategy.

toolkits for expeditionary forensic activities.

- Africa Command issued guidance to support requests for conducting forensic activities on behalf of foreign partners in Africa.

- European Command published guidance on using forensic capabilities to support operations in Europe.

Additionally, the DOD forensic directive established a Forensic Executive Committee,[19] in 2011, to, among other things, facilitate the coordination of forensic activities throughout DOD. At the time of our review, the Forensic Executive Committee had not met because, according to an OUSD(AT&L) official, there had not been any significant issues on forensic activities that required the attention of OUSD(AT&L) senior leadership. OUSD(AT&L) created a Coordination Steering Group in 2011 as a forum for DOD components to collectively identify and prioritize issues, analyze alternatives, and develop recommendations for approval by the Forensic Executive Committee. The Coordination Steering Group has created working groups to examine performance metrics, research and development efforts, and policy related to maintaining an enduring expeditionary forensic capability. OUSD(AT&L) officials stated that the Coordination Steering Group has reviewed the draft strategic plan and provided its comments. Nonetheless, the lack of an approved strategic plan, and associated implementation plans, limits DOD's ability to effectively and efficiently prioritize its efforts to develop an enduring expeditionary forensic capability by the end of 2014.

OUSD(AT&L) Has Not Reviewed and Evaluated the Adequacy of Expeditionary Forensic Budget Estimates for Fiscal Years 2013 through 2018

OUSD(AT&L) has not reviewed and evaluated the adequacy of the DOD components' expeditionary forensic budget estimates for fiscal years 2013 through 2018. As required by the directive, USD(AT&L) is to review the adequacy of forensic-related acquisition programs and associated budget submissions to ensure they meet the Defense Forensic Enterprise requirements and objectives. At the time of our review, OUSD(AT&L) officials stated that they had most recently reviewed the DOD components' fiscal year 2012 forensic budget estimates, which includes

---

[19]The Forensic Executive Committee shall be chaired by OUSD(AT&L) or a designated representative, and consists of one senior general or flag officer or civilian equivalent from each of the DOD components that provide or require forensic capabilities and support, as well as those making strategic management decisions related to forensic activities.

expeditionary forensic budget data. In addition, OUSD(AT&L) officials said that they had requested the DOD components' forensic budget estimates for fiscal year 2013, but had not reviewed and evaluated these estimates to ensure they meet the Defense Forensic Enterprise requirements and objectives because the department was operating under a continuing resolution and therefore was adhering to fiscal year 2012 budget levels. However, OUSD(AT&L) officials stated that they had not reviewed and evaluated the adequacy of the DOD components' expeditionary proposed forensic budget estimates for fiscal years 2014 through 2018, and had not issued a data call to obtain this information from the DOD components until our formal request in September 2012 in order to conduct our review. Based on the data from our formal request, OUSD(AT&L) officials estimated that the DOD components need about $363.5 million to fund expeditionary forensic capabilities from fiscal years 2013 through 2018; however, this conclusion was drawn without OUSD(AT&L) evaluating the data. OUSD(AT&L) officials said that they had not reviewed and evaluated DOD components' forensic budget estimates because, among other things, OUSD(AT&L) was waiting for the DOD components to finalize their proposed budget estimates for fiscal years 2014 through 2018, and waiting for the Joint Capabilities Integration and Development System to validate their forensic requirements. Regardless of not having validated forensic requirements, reviewing and evaluating the DOD components' proposed budget estimates allows OUSD(AT&L) to advise the DOD components on their resource allocation decisions with respect to expeditionary forensic capabilities and ascertain whether the proposed funding is adequate to meet the department's overarching requirements and objectives. Moreover, DOD officials have noted the need to establish base funding for expeditionary forensic capabilities in advance of expected reductions in Overseas Contingency Operations funding. Given the competition for limited resources within their base budgets, DOD officials said that if the department does not take proactive measures to reprogram funds for expeditionary forensic activities from Overseas Contingency Operations to base budget accounts, the military services will experience a significant funding gap that could limit their ability to respond to current and future warfighting requirements.

OUSD(AT&L) officials cited several factors that also affected their ability to review and evaluate the DOD components' forensic budget data, including expeditionary forensics. For example, the DOD components' budget estimates for expeditionary forensics are often aggregated with other costs and distributed across numerous budget accounts that are not explicitly identified as forensic activities. As a result, these officials noted the difficulty in being able to identify forensic activities within operation and maintenance accounts. In addition, these officials cited issues in determining which types of forensic-related costs to identify, such as training, research and development, and information systems. We also identified similar issues concerning the reliability of the forensic budget data.

OUSD(AT&L) officials said that if DOD components were instructed to collect budget data on forensic activities, such as Expeditionary Forensic Laboratories, training, personnel, and research and development, then OUSD(AT&L) would be better positioned to review and evaluate their forensic budget estimates. However, OUSD(AT&L) said the DOD components do not have guidance on collecting and reporting their expeditionary forensic activities. Furthermore, the DOD directive does not include instructions on collecting and reporting forensic budget data. According to GAO's *Standards for Internal Control in the Federal Government*, agencies should provide policy and guidance to determine the effectiveness and efficiency of operations, including the use of resources needed to achieve their goals.[20] Without collection and reporting guidance, OUSD(AT&L) will continue to experience challenges with reviewing and evaluating the costs associated with DOD's expeditionary forensic capabilities.

Conclusions

DOD recognizes the value of expeditionary forensics for identifying and targeting enemy combatants and terrorists, and has taken actions to establish expeditionary forensics as an enduring capability across the full range of military operations by the end of 2014. To achieve a coordinated, holistic approach across the department, DOD has issued a forensic directive that requires, among other things, USD(AT&L) to publish a strategic plan and review the adequacy of forensic-related acquisition programs and associated budget submissions. However, the strategic plan has been in draft for 2 years with no publication date set—and by extension, no publication date has been set for the proposed implementation plans. Further, without key elements, such as approaches, milestones and metrics, and identification of needed resources, DOD will be unable to measure its progress and adjust its approach and resourcing as necessary to achieve the goals and

---

[20] GAO/AIMD-00-21.3.1.

objectives outlined in the forensic strategic plan. In the absence of an approved strategic plan, several military services and combatant commands have been independently developing their own guidance and plans to inform their specific forensic activities. Further, because USD(AT&L) has not reviewed and evaluated the adequacy of the DOD components' expeditionary forensic budget estimates, there is no assurance that these proposed budget estimates are consistent with one another and with the department's overarching goals and objectives. As a result, DOD's ability to fund its expeditionary forensic requirements in the most efficient and effective manner may be adversely affected.

Recommendations for Executive Action

As DOD establishes an enduring expeditionary forensic capability prior to the projected drawdown of operations in Afghanistan by the end of 2014, we recommend that the Secretary of Defense direct the Under Secretary of Defense for Acquisition, Technology, and Logistics to take the following four actions:

- Incorporate key elements in its forensic strategic plan, implementation plans, and other associated guidance that are currently absent including approaches for achieving goals and objectives, milestones and metrics to gauge the department's progress, and resources needed to meet its goals and objectives.

- Set a date to publish the strategic plan for the Defense Forensic Enterprise.

- Periodically review and evaluate the DOD components' proposed forensic budget estimates—including expeditionary forensics—to help ensure the department's overarching requirements and objectives will be met, in accordance with the DOD Defense Forensic Enterprise directive.

- Issue guidance on how DOD components are to collect and report their forensic budget data—including expeditionary forensic budget data.

Agency Comments and Our Evaluation

In written comments on this draft, DOD agreed with all four of our recommendations and discussed steps it plans to take to address these recommendations. DOD's written comments are reprinted in their entirety in appendix II. DOD also provided technical comments, which we have incorporated into the report where appropriate.

DOD concurred with our first recommendation that the Secretary of Defense direct the Under Secretary of Defense for Acquisition, Technology, and Logistics to incorporate key elements which are currently absent in its forensic strategic plan, implementation plans and other associated guidance. DOD stated that upon approval of the Defense Forensic Enterprise Strategic Plan, it will begin drafting the forensic strategy's implementation plan to include specific priorities and tasks that support the goals and objectives of the strategic plan. The forensics implementation plan also will assign an office of primary responsibility to accomplish each task, propose milestones, develop success metrics, and estimate required resources.

DOD concurred with our second recommendation to the Secretary of Defense to direct the Under Secretary of Defense for Acquisition, Technology, and Logistics to set a date to publish the strategic plan for the Defense Forensic Enterprise and stated that it anticipates publishing the Defense Forensic Enterprise Strategic Plan prior to the end of fiscal year 2013.

DOD concurred with our third recommendation to the Secretary of Defense to direct the Under Secretary of Defense for Acquisition, Technology, and Logistics to periodically review and evaluate the DOD components' proposed forensic budget estimates—including expeditionary forensics—to help ensure the department's overarching requirements and objectives will be met in accordance with the DOD Defense Forensic Enterprise directive. DOD said that the Under Secretary, Comptroller, the Joint Staff, and the military services will coordinate to identify mechanisms to more efficiently and reliably review DOD components' proposed forensic budget estimates against validated requirements. While we are encouraged with DOD's decision, we believe that it is important that DOD schedule its reviews at regular intervals, in accordance with the Defense Forensic Enterprise directive.

DOD concurred with our fourth recommendation to the Secretary of Defense to direct the Under Secretary of Defense for Acquisition, Technology, and Logistics to issue guidance on how DOD components are to collect and report their forensic budget data—including expeditionary forensic budget data. DOD stated that the Under Secretary, Comptroller, and the military services will coordinate to develop and issue guidance on reporting forensic budget data prior to the beginning of the fiscal year 2016 budget planning cycle.

We are sending copies of this report to appropriate congressional committees; the Secretary of Defense; the Under Secretary of Defense for Acquisition, Technology and Logistics; the Chairman, Joint Chiefs of Staff; the Secretaries of the Army, the Navy, and the Air Force; the Commandant of the Marine Corps; and the Director, Office of Management and Budget. In addition, this report will be available at no charge on the GAO Website at http://www.gao.gov.

If you or your staff have any questions about this report, please contact me at (202) 512-4523 or at leporeb@gao.gov. Contact points for our Office of Congressional Relations and Public Affairs may be found on the last page of this report. Key contributors to this report are listed in appendix III.

Brian J. Lepore
Director
Defense Capabilities and Management

List of Requesters

The Honorable Adam Smith
Ranking Member
Committee on Armed Services
House of Representatives

The Honorable Mac Thornberry
Chairman
The Honorable Jim Langevin
Ranking Member
Subcommittee on Intelligence, Emerging Threats and Capabilities
House of Representatives

The Honorable Jeff Miller
House of Representatives

Appendix I: Scope and Methodology

To assess the steps DOD has taken to establish an enduring expeditionary forensic capability, we analyzed relevant policies and guidance from the Office of the Secretary of Defense, such as the Department of Defense's (DOD) forensic directive,[1] draft strategic plan for the Defense Forensic Enterprise,[2] and the capstone concept of operations for forensics.[3] In addition, we analyzed relevant Army, Marine Corps, and combatant command plans, policies and guidance that describe DOD's efforts to organize, train, and equip forces to carry out expeditionary forensic activities. To understand DOD's current expeditionary forensic capabilities, including those of the Expeditionary Forensic Laboratories, we visited the Army's Criminal Investigation Laboratory at Fort Gillem, Georgia, which is primarily responsible for conducting both traditional and expeditionary forensic activities for the department and is known as DOD's forensic science center of excellence. We also met with officials from CNA's Center for Naval Analyses to discuss their efforts to review DOD's overall defense forensic activities.[4] Further, we interviewed officials from the DOD organizations identified in table 2 to discuss DOD's efforts to establish an enduring expeditionary forensic capability.

Table 2: DOD Organizations Contacted[a]

Office of the Secretary of Defense	Office of the Under Secretary of Defense (Comptroller)
	Office of the Under Secretary of Defense for Policy, Defense Prisoner of War/Missing Personnel Office
	Office of the Under Secretary of Defense for Intelligence
	Defense Intelligence Agency
	Office of the Under Secretary of Defense for Acquisition, Technology, and Logistics; Office of the Assistant Secretary of Defense, Research and Engineering; Defense Biometrics & Forensics

[1]Department of Defense, *DOD Forensic Enterprise,* DOD Directive 5205.15E (April 26, 2011).

[2]*DOD Forensic Enterprise Draft Strategic Plan* (as of February 2013).

[3]DOD *Capstone Concept of Operations for Department of Defense Forensics Enterprise* (Jul. 18, 2008).

[4]CNA is a not-for profit research and analysis organization and a Federally Funded Research and Development Center. CNA is the parent organization of the Institute for Public Research and the Center for Naval Analyses.

Appendix I: Scope and Methodology

The Joint Staff	Force Structure, Resources, and Assessment Directorate; Requirements, Force Protection Division
U.S. Army	Office of the Provost Marshal General
	Criminal Investigation Command
	Army Criminal Investigation Laboratory, Fort Gillem, Georgia
	Biometrics Identity Management Agency
	Training and Doctrine Command Capabilities Manager – Biometrics & Forensics, Fort Huachuca, Arizona
	Headquarters, Office of Operations and Plans, Capabilities and Integration Command
	Headquarters, Office of Operations and Plans, Force Management
U.S. Navy	Office of the Chief of Naval Operations, Anti-Terrorism, Force Protection, Chemical, Biological, Radiological, and Nuclear Office, Surface Warfare Directorate
	Office of the Chief of Naval Operations, Deputy Chief of Naval Operations for Information Dominance Operational Integration and Capabilities
	Office of the Chief of Naval Operations, Expeditionary Warfare Division, Navy Expeditionary Combat Branch
	Naval Sea Systems Command, Anti-Terrorism Afloat Office
	Naval Criminal Investigative Service, Biometrics Division
U.S. Marine Corps	Headquarters, Plans, Policies, and Operations Division, Identity Operations Section
U.S. Air Force	Defense Cyber Crime Center
	Office of Special Investigations
U.S. European Command	Directorate of Intelligence, Strategy Division, Identity Intelligence Branch, Stuttgart, Germany
U.S. Southern Command	Identity Intelligence Program, Miami, Florida
U.S. Special Operations Command	Identity Operations Program, MacDill Air Force Base, Florida
U.S. Northern Command	Risk Management Branch, Force Protection and Mission Assurance Division, Peterson Air Force Base, Colorado
U.S. Central Command	Joint Security Office Joint Security Force Protection Technology Branch, MacDill Air Force Base, Florida
U.S. Pacific Command	Forensic Exploitation Division, Camp H.M. Smith, Hawaii
U.S. Africa Command	Identity Resolution Program, Stuttgart, Germany

Source: GAO.

[a]Unless otherwise indicated, these organizations are located within the Washington, D.C., metropolitan area.

We also reviewed and analyzed the military services' and Special Operations Command's current and projected forensic budget estimates to determine if the data were sufficiently reliable and met the department's requirements in the DOD forensic directive. We assessed the reliability of this budget data by interviewing knowledgeable officials and reviewing related documentation and written responses to our questions on data reliability. We identified several issues concerning the reliability of the budget data obtained from Office of the Under Secretary of Defense for Acquisition, Technology, and Logistics (OUSD(AT&L)), the

Appendix I: Scope and Methodology

military services and Special Operations Command, including the sources from which the data were derived, the consistency in how the data were compiled, and the manner in which the data were verified. As a result, we determined that the data were not sufficiently reliable. Therefore, we are making a recommendation that addresses OUSD(AT&L)'s ability to review and evaluate the DOD components' forensic budget data by calling for the development of budget collecting and reporting guidance.

We interviewed officials from the Department of Justice, Federal Bureau of Investigation, and Department of Homeland Security to obtain their perspective on DOD's efforts to develop an expeditionary forensic capability. We also met with officials from the National Science and Technology Council within the Executive Office of the President to gain an understanding of national policy trends on forensics.

We conducted this performance audit from May 2012 through June 2013 in accordance with generally accepted government auditing standards. Those standards require that we plan and perform the audit to obtain sufficient, appropriate evidence to provide a reasonable basis for our findings and conclusions based on our audit objectives. We believe that the evidence obtained provides a reasonable basis for our findings and conclusions based on our audit objectives.

Appendix II: Comments from the Department of Defense

ASSISTANT SECRETARY OF DEFENSE
3030 DEFENSE PENTAGON
WASHINGTON, DC 20301-3030

RESEARCH
AND ENGINEERING

JUN 20 2013

Mr. Brian J. Lepore
Director, Defense Capabilities and Management
U.S. Government Accountability Office
441 G Street, N.W.
Washington, DC 20548

Dear Mr. Lepore:

This is the Department of Defense (DoD) response to Government Accountability Office (GAO) Draft Report, GAO-13-447, "Defense Forensics: Additional Planning and Oversight Needed to Establish an Enduring Expeditionary Forensic Capability," dated May 13, 2013 (GAO 351729). Detailed comments on the report recommendations are enclosed.

The Department has taken important steps to establish an enduring expeditionary forensic capability and recognizes that more work is needed. The Department will continue to work within the Joint Capabilities Integration Development and the Planning, Programming, Budgeting, and Execution systems to balance resources against all requirements including forensics.

Sincerely,

Alan R. Shaffer
Acting

Enclosure:
Response to GAO Draft Report, GAO-13-447 (GAO Code 351729)

Appendix II: Comments from the Department of Defense

GAO Draft Report Dated May 13, 2013
GAO-13-447 (GAO CODE 351729)

"DEFENSE FORENSICS: ADDITIONAL PLANNING AND OVERSIGHT NEEDED TO ESTABLISH AN ENDURING EXPEDITIONARY FORENSIC CAPABILITY"

DEPARTMENT OF DEFENSE COMMENTS
TO THE GAO RECOMMENDATION

RECOMMENDATION 1: The GAO recommends that the Secretary of Defense direct the Under Secretary for Acquisition, Technology and Logistics to take action to incorporate key elements in its forensic strategic plan, implementation plans and other associated guidance that are currently absent including approaches for achieving goals and objectives, milestones and metrics to gauge the department's progress and resources needed to meet its goals and objectives.

DoD RESPONSE: Concur. Upon approval of the Defense Forensic Enterprise Strategic Plan, the Office of the Under Secretary for Acquisition, Technology and Logistics will begin drafting the forensic strategy's implementation plan. The forensics implementation plan will identify specific priorities and tasks that support the goals and objectives of the strategic plan. The forensics implementation plan will also assign an office of primary responsibility to accomplish each task, propose milestones, develop success metrics and estimate required resources.

RECOMMENDATION 2: The GAO recommends that the Secretary of Defense direct the Under Secretary for Acquisition, Technology and Logistics to take action to set a date to publish the strategic plan for the Defense Forensic Enterprise.

DoD RESPONSE: Concur. The Office of the Under Secretary for Acquisition, Technology and Logistics anticipates publishing the Defense Forensic Enterprise Strategic Plan prior to the end of fiscal year 2013.

RECOMMENDATION 3: The GAO recommends that the Secretary of Defense direct the Under Secretary for Acquisition, Technology and Logistics to take action to periodically review and evaluate the DoD components' proposed forensic budget estimates-including expeditionary forensics-to help ensure the department's overarching requirements and objectives will be met, in accordance with the DoD Defense Forensic Enterprise directive.

DoD RESPONSE: Concur. The Office of the Under Secretary for Acquisition, Technology and Logistics will coordinate with the DoD Comptroller, the Joint Staff and the Military Services to identify mechanisms to more efficiently and reliably review DoD Components' proposed forensic budget estimates against validated requirements.

RECOMMENDATION 4: The GAO recommends that the Secretary of Defense direct the Under Secretary for Acquisition, Technology and Logistics to take action to issue guidance on how DoD components are to collect and report their forensic budget data-including expeditionary forensic budget data.

Appendix II: Comments from the Department of Defense

DoD RESPONSE: Concur. The Office of the Under Secretary for Acquisition, Technology and Logistics will coordinate with the DoD Comptroller and Military Services to develop and issue guidance on reporting forensic budget data prior to the beginning of the fiscal year 2016 budget planning cycle.

Appendix III: GAO Contact and Staff Acknowledgments

GAO Contact	Brian J. Lepore, Director, 202-512-4523 or leporeb@gao.gov
Staff Acknowledgments	In addition to the contact named above, Marc Schwartz, Assistant Director; Grace Coleman; Latrealle Lee; Alberto Leff; Amber Lopez Roberts; Tim Persons; Terry Richardson; Amie Steele; Sabrina Streagle; John Van Schaik; and Nicole Willems made key contributions to this report.

Related GAO Products

Afghanistan: Key Oversight Issues, GAO-13-218SP. Washington, D.C.: February 11, 2013.

Counter-Improvised Explosive Devices: Multiple DOD Organizations Are Developing Numerous Initiatives. GAO-12-861R. Washington, D.C.: August 1, 2012.

Urgent Warfighter Needs: Opportunities Exist to Expedite Development of Fielding of Joint Capabilities. GAO-12-385. Washington, D.C.: April 24, 2012.

Defense Biometrics: Additional Training for Leaders and More Timely Transmission of Data Could Enhance the Use of Biometrics in Afghanistan. GAO-12-442. Washington, D.C.: April 23, 2012.

Defense Biometrics: DOD Can Better Conform to Standards and Share Biometric Information with Federal Agencies. GAO-11-276. Washington, D.C.: March 31, 2011.

Warfighter Support: Actions Needed to Improve Visibility and Coordination of DOD's Counter-Improvised Explosive Device Efforts. GAO-10-95. Washington, D.C.: October 29, 2009.

Defense Management: DOD Can Establish More Guidance for Biometrics Collection and Explore Broader Data Sharing. GAO-09-49. Washington, D.C.: October 15, 2008.

Defense Management: DOD Needs to Establish Clear Goals and Objectives, Guidance, and a Designated Budget to Manage Its Biometrics Activities. GAO-08-1065. Washington, D.C: September 26, 2008.

GAO's Mission	The Government Accountability Office, the audit, evaluation, and investigative arm of Congress, exists to support Congress in meeting its constitutional responsibilities and to help improve the performance and accountability of the federal government for the American people. GAO examines the use of public funds; evaluates federal programs and policies; and provides analyses, recommendations, and other assistance to help Congress make informed oversight, policy, and funding decisions. GAO's commitment to good government is reflected in its core values of accountability, integrity, and reliability.
Obtaining Copies of GAO Reports and Testimony	The fastest and easiest way to obtain copies of GAO documents at no cost is through GAO's website (http://www.gao.gov). Each weekday afternoon, GAO posts on its website newly released reports, testimony, and correspondence. To have GAO e-mail you a list of newly posted products, go to http://www.gao.gov and select "E-mail Updates."
Order by Phone	The price of each GAO publication reflects GAO's actual cost of production and distribution and depends on the number of pages in the publication and whether the publication is printed in color or black and white. Pricing and ordering information is posted on GAO's website, http://www.gao.gov/ordering.htm.
	Place orders by calling (202) 512-6000, toll free (866) 801-7077, or TDD (202) 512-2537.
	Orders may be paid for using American Express, Discover Card, MasterCard, Visa, check, or money order. Call for additional information.
Connect with GAO	Connect with GAO on Facebook, Flickr, Twitter, and YouTube. Subscribe to our RSS Feeds or E-mail Updates. Listen to our Podcasts. Visit GAO on the web at www.gao.gov.
To Report Fraud, Waste, and Abuse in Federal Programs	Contact: Website: http://www.gao.gov/fraudnet/fraudnet.htm E-mail: fraudnet@gao.gov Automated answering system: (800) 424-5454 or (202) 512-7470
Congressional Relations	Katherine Siggerud, Managing Director, siggerudk@gao.gov, (202) 512-4400, U.S. Government Accountability Office, 441 G Street NW, Room 7125, Washington, DC 20548
Public Affairs	Chuck Young, Managing Director, youngc1@gao.gov, (202) 512-4800 U.S. Government Accountability Office, 441 G Street NW, Room 7149 Washington, DC 20548

Please Print on Recycled Paper.

www.ingramcontent.com/pod-product-compliance
Lightning Source LLC
Chambersburg PA
CBHW081812170526
45167CB00008B/3407